Date Due

How to Overcome Fear of the Dentist

A Patient's Guide to Understanding Dentistry

Bertrand Bonnick D.D.S.,
M.A.G.D., D.D.O.C.S., A.F.A.A.I.D.
& Kaye Bonnick, MBA

AuthorHouse™
1663 Liberty Drive
Bloomington, IN 47403
www.authorhouse.com
Phone: 1-800-839-8640

First published by AuthorHouse 11/2/2010

ISBN: 978-1-4490-7057-1 (sc)
ISBN: 978-1-4490-7058-8 (e)

Library of Congress Control Number: 2010908369

Printed in the United States of America
Bloomington, Indiana

This book is printed on acid-free paper.

This book is dedicated to my dental team, and patients, who made me the dentist I am today.

Special thanks to Kaye Bonnick who spent many hours refining my thoughts and designing this book.

Contents

Foreword

I first met Dr. Bonnick in Frankfurt, Germany as VIP guests of a major dental manufacturer. They were updating us on their research in several implant systems as well as the new development and success of one particular system. Although I practice in California and he is on the East Coast, we still find time to stay in touch and share ideas. As a practicing oral surgeon, I am always interested in the proper treatment, planning, and execution of the restorative aspects of my implant placement, as well as how the restoring dentist planned to maintain his patient's oral rehabilitation.

After reading his manuscript, I am reminded of ideas we have discussed in an effort to explore better ways of treating our patients. Why should you read this book? Why should you have confidence in Dr. Bonnick? In the quest for good oral health care, the best doctor for the job is the one who is committed to staying abreast of current developments and the one who continues to learn about advanced techniques that will benefit patients.

Dr. Bonnick has continued to pursue his education in all the major areas of dentistry and was recognized as a Master in the Academy of General Dentistry in 2006. While it's great to have advanced dental techniques to solve your problems, wouldn't you feel better knowing that you could have that treatment without fear or pain?

Those who are fortunate enough to be a patient of Dr. Bonnick will receive the highest quality of care from one of the top doctors in the field of general dentistry and implant surgery. As a reader of this book, you will receive the information he uses to make treatment decisions on a daily basis, as well as plan for your dental challenges. Dr. Bonnick has trained nationally and internationally and has mastered techniques of cosmetic, sedation, and implant dentistry. He is also an instructor of the latest techniques in implant and cosmetic dentistry. Dr. Bonnick offers courses in his office for dental colleagues and their staff every year, as well as lecturing at various dental meetings. You will learn the value of good oral care and receive education on how important it is for a foundation of oral health to begin at an early age. You will also learn about oral health care as it relates to overall medical health and well-being.

So smile and be happy, because through the covers of this book, you will see how Dr. Bonnick is not only a smart and talented doctor, but is also compassionate and committed to teaching you how to improve your dental health.

Dr. Bonnick is highly motivated to help you overcome your dental shortcomings, and points you in the direction that restores you to optimal oral health and beauty. He will help you alleviate your dental fears through education and knowledge of sedation dentistry. He will encourage you to rejuvenate your smile with the most advanced of technologies in the field of dentistry today. Enjoy the literary journey.

Dr. Arthur Johnson
Diplomate American Association of Oral and Maxillofacial Surgeons
Laguna Niguel, California, USA

Preface

The key to overcoming dental fear is to become educated about dental processes and knowing that options are available to make dental procedures less stressful for fearful patients. Don't delay dental care because you are scared. We are living in a day of advanced technology and techniques that addresses this real problem for so many patients, including me.

At age fifteen, I sat at a table over a plate of my favorite meal; the smell heightened my anticipation of what was to come. It was my favorite meal, a mixture of sautéed codfish spiced with the exotic sections of golden ackees slowly cooked to perfection. I was so hungry I could anticipate the wonderful taste of the food from its smell. I took a bite and started to chew. The food was delicious. Suddenly, and without warning, a pain like a lightning strike violently shot through my head and tears came to my eyes. The food had lodged in a large hole in my tooth, and the pain was unbearable. I could not eat the rest of my meal, and I went to bed hungry and in pain—continuous throbbing pain.

I grew up in an area with non-fluoridated water, and being a "macho male," I thought flossing was something you did when meat got caught between your teeth. I was even skeptical about getting my teeth cleaned, because I thought the dentist would remove some of my enamel and make my teeth weaker.

I was experiencing constant pain and had no choice but to go to the dentist. I went, and he declared that I had eleven cavities and he would be happy to "pull" them all. He advised me to come see him every Friday and he would extract a tooth each week until the job was done.

He pulled one of my back teeth every week, and on the third visit, I knew I would not return. I did not go back to that dentist, but he made an indelible impression on me. I remember him saying, "You will only feel a little pinch," and "Be a man!" It was a scary experience, but the impression it made on my mind stayed with me for a lifetime. I was scared of the dentist and the dental office.

Three weeks of dental trauma created a lifetime of dental phobia for me, but it also determined my career. I decided that I would become a dentist and I would find a better way. It also made me an almost obsessive-compulsive brusher, and that helped me reduce the amount of cavities throughout my teen years. Dental cavities were under control, but at age twenty-two, I was diagnosed with early periodontal disease in my first year of dental school. I went through five sessions with the hygienist to control the inflammation, and twenty-seven years later, I still have all of my remaining teeth. Since graduating from dental school in 1986, I have been "preaching the religion of preventive dentistry" and finding ways to help dental-phobic patients cope with their fears.

Ringing in my ears over all these years were the words I muttered during and after my extraction dental visits: "There must be a better way!" In retrospect, I would like to thank the dentist who gave me a case of dental phobia and fueled my desire to come up with a better way. It has led me to zealously pursue continuing dental education in order to find a better way.

I have been on several mission trips to third world countries; I did two years of general-practice hospital dentistry. I taught in a hospital dental program part time for ten years, worked in a prison for a year and a half, attended a one-year Maxicourse on implant dentistry, and pursued credentials with three dental organizations. I still strive to do more. I wish to take a moment now to explain what the letters behind my name indicate.

M.A.G.D.—Mastership in the Academy of General Dentistry requires at least 1,100 hours of continuing education, including a written examination after the first 500 hours, which gives you the status of Fellowship. Upon completion of the next 600 hours—400 of which are participation courses—you are awarded Mastership.

A.F.A.A.I.D.—Associate Fellow in the American Academy of Implant Dentistry required passing a written exam, recording successful implant cases of various types and presenting them to selected members of the Academy followed by a verbal cross examination and testing that you have to pass to their satisfaction.

D.D.O.C.S. - Diplomate in the Dental Organization for Conscious Sedation required writing numerous articles, recording several hundred cases of sedated patients, passing exams in Advanced Cardiac Life Support, professional recommendations, taking numerous classes, and passing an oral examination. This credential is open to those members who had already achieved the status of Fellowship in the organization.

I do not consider myself to be a better dentist than those who do not have my credentials, but I know that I am a better dentist than I would have been had I not pursued additional training in order to find "a better way."

1

Sedation Dentistry

When I was fifteen years old, I had eleven cavities. I went to a dentist, who extracted three of my teeth and would have taken out more if I was cooperative. He told me, "You will only feel a little pinch." To him it was a pinch; to me it was a terrifying ordeal, and I vowed if I could ever improve the process, I would.

Since the late 1800s, dentists have always used various methods of pain control in order to allow their patients to relax and have their work done. Some patients could have an extraction with local anesthesia but needed additional levels of sedation for a root canal. Other patients could have a root canal with local but needed sedation for wisdom tooth extractions. It is all about perception. Some patients are scared when they hear the sound of a drill; others are bothered by the sight of a needle or the smell of medicine they associate with a dental office. Some people have a gag reflex—some so bad that they start to gag when a hand with a dental instrument approaches their face. Dental fear is an individual thing.

Based on a patient's health and the dental procedures needed, we can propose different levels of sedation. A patient treated at one level for one procedure may be treated with less for a similar procedure at a different time. For purposes of discussion for general dentistry, we can call it "the five levels of sedation."

The Five Levels of Sedation

Level One: Iatro-sedation or techniques the doctor can use to make the patient feel more comfortable. This may include hypnosis; use of hard- and soft-tissue lasers, warming of the local anesthetics; or using small pediatric needles to start all injections; speaking to the patient in a low, calm voice.

Level Two: Adding nitrous oxide to decrease the patient's apprehension along with local anesthesia as needed, and other techniques used in Level One.

Level Three: Anti-anxiety or relaxation medications are given to patients intra-orally (by mouth) prior to treatment. Patients go through treatment in a more relaxed state. Some amnesic properties are present in these medications. Patients pulse, heart rate, and blood pressure are monitored.

Level Four: Treatment is done with the addition of intravenous (by vein) sedation medication to produce a mild conscious sedation state. The patient can maintain his or her own airway and can respond to stimulus, thus minimizing the possibility of low oxygen levels and cardiac events. In addition to the patients pulse, heart rate, and blood pressure being monitored, a capnography machine is also used to monitor their carbon dioxide levels.

Level Five: Treatment is done under general anesthesia in a hospital setting, monitored by an anesthesiologist.

If you are hesitant to do dental treatment or even visit a dentist, try discussing these levels of sedation with your personal dentist. There is always an option that can help you.

Keeping Sedation Safe with Good Monitoring Equipment

Capnography allows us to see a graph of the patients' carbon dioxide level as they breathe. This is one of the best additions to in-office sedation dentistry that I have ever used. This device allows the doctor to monitor the sedated patients' carbon dioxide levels to ensure that the right levels are maintained.

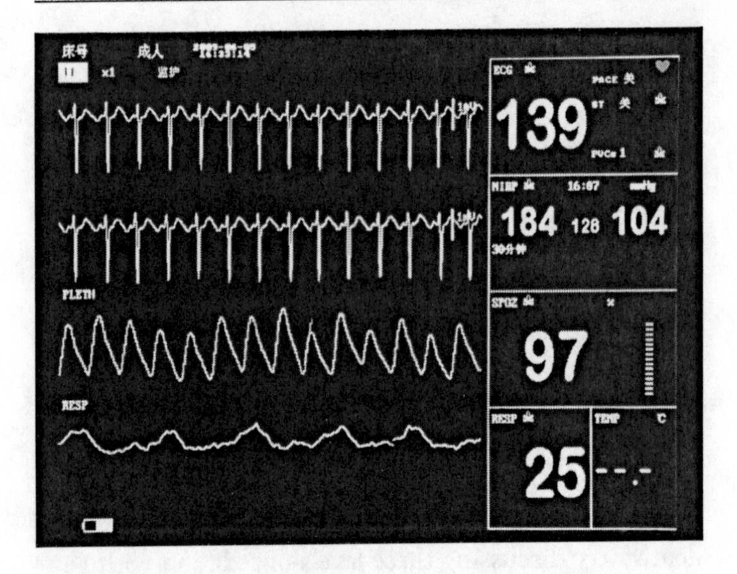

Another great device used in sedation dentistry is an EKG, which can give us a good idea of your basic heart rhythms during a moderate sedation session. EKG, blood oxygen level, blood pressure, pulse rate, and carbon dioxide levels, are parameters that quickly alert us to the level of sedation, as well as alertness, and provide us with a way of continually monitoring the patient through a dental visit.

One of the best safety measures in sedation dentistry is the accuracy of the patient's medical history. Accurate history includes past use of drugs, including illicit drugs, operations, medications, history of illnesses, and recent medical checkups. Sedation dentistry involves a whole team, and the team consists of not only the people giving care but the patient and their full involvement with their treatment.

Undiagnosed or untold medical conditions are one of the greatest hindrances to proper treatment of a patient undergoing sedation dentistry. Do not keep parts of your medical history a secret from your sedation dentist!

2

Control Your Fears—Control Your Costs

Dentistry is one of the areas of healthcare that responds readily to prevention. While many in the population have little dental fear, our culture and media are replete with images of dentistry indicating that dental treatment is bothersome for most people. Phrases like "It was worse than a root canal" or "He is not going to hurt you; he is just going to look" or "You will only feel a little pinch" are common. Jokes abound with references to numb lips and tongue, dentists putting their knees on people's chest to aid in the extraction of teeth, as well as women who state that they would rather have a child than have their teeth worked on. So how soon should you start taking care of your teeth?

The recommendation of pediatric dentists is that the first visit for your child should occur around his or her first birthday. Signs of fluorosis (excess fluoride) in the teeth, tongue and finger-sucking habits, as well as nursing bottle caries (widespread decay on several teeth as a result of children nursing without follow-up oral hygiene), can be intercepted and lead to better long-term results for the child and less expenses for the parents. Early intervention, such as placement of sealants to prevent cavities, can save a lot of costs in the long run. Better brushing and flossing habits, as well as the application of topical fluoride, could enhance a lifetime of dental health.

Some parents still believe "If it is a baby tooth, pull it; if it is a permanent tooth, save it!" In adult life, the mantra is "If it is a front tooth, save it; if it is a back tooth, pull it!" Both beliefs lead to extremely bad results for patients and increase the cost for replacing teeth in the proper position. The amount of trauma and fear that is imprinted on people during these dental experiences adds to the mass of people who fear the dentist.

Baby teeth (primary or deciduous teeth) are very important in preserving the space for the eruption of the adult teeth (secondary teeth or permanent teeth). They are important because early tooth loss could lead to ill-shaped and wrongly positioned teeth, which would require braces to place the teeth in the right position. Later in life, the increased need for crown and bridge restorations to replace and preserve teeth costs more.

It is not always easy to differentiate between baby teeth and adult teeth. Radiographs (X-ray representation) of teeth will also reveal that some primary teeth do not have secondary successors, and many adults retain primary teeth throughout their lives. Early loss of baby teeth can lead to the non-eruption of adult teeth or their eruption in abnormal positions.

Abnormally positioned teeth are harder to maintain and do not function as well as properly positioned teeth. Abnormally positioned teeth produce a malocclusion (teeth coming together in an unfavorable way) that can contribute to temporomandibular joint disorder (TMD) and the inability to properly clean areas between these teeth. This creates a breeding ground for bacteria that helps to promote disease of the gums and the supporting bone for the teeth (alveolar bone). Teeth that do not fit together properly contribute to conditions that increase the cost of dental care.

Back teeth were designed to bear the pressure of the clenching jaws, while front teeth are designed for incising or cutting. Partially erupted, unerupted, or erupted wisdom teeth in bad positions may need to be extracted as soon as possible. Your dentist can help you in making that determination. The other back teeth crush food and also keep the chewing muscles in shape, as well as supporting the structures of the face. The better shape the facial muscles are in, the less collapsed the appearance of the face. Extracting one of these back teeth leads to a slow drifting of the teeth on either side, as well as movement of the opposing tooth toward the spot where the missing partner tooth was. These movements lead to widening spaces and/or slanting teeth, as well as less ability to chew food properly. Eventually, loss of enough back teeth without replacement means more pressure will be on the front teeth that are not designed for this purpose. Spreading of the front teeth and/or collapse of the opposing teeth leads to a collapsed lower face that is often equated with faster aging.

Correcting early baby teeth loss can save thousands of dollars in braces, as well as bridgework and implants. Keeping back teeth by doing procedures like root canals followed by crowns is very important to preventing collapse of the lower face. Procedures such as extractions, followed by bone grafting and implant placement, help to preserve the integrity of the oral structures. Later in the book, we will show how spending a little more up front can save major rehabilitative expenses later on.

Fear of the dentist in any form can prevent patients from taking advantage of the many new developments in dental care. This can create savings that result from doing small procedures instead of large procedures. One filling, to stop a cavity from getting into the nerve, costs approximately one-tenth the cost to do a root canal, post and core and crown, restoring the tooth once the decay got into the nerve and infected it.

Recent advancement in sedation dentistry allows most people to have their necessary dental work done. The more involved the dental work, the more costly it will be. It is up to patients to take some responsibility in finding a sedation dentist if fear is what keeps them from having their dental work done.

If patients practiced effective prevention, they would only need major procedures by choice, genetics, wear and tear of the dentition, or as a result of trauma. Microbiologic studies indicate that the bacterial buildup that leads to plaque and calculus development takes place over the course of three months. If you are really serious about prevention, you should have a preventive visit to the dentist every three months.

In the following pages, you will learn ways of safeguarding you and your family's dental health in such a way that you will spend less in the long run on dental care. For people who have long since lost the battle to preserve their teeth and supporting bone, there is hope. Extensive research and procedures for growing and grafting new bone, as well as three-dimensional imaging systems, help dentists simulate existing conditions and plan the rebuilding of the mouth and supporting structures.

3

Building a Good Foundation for Good Dental Health

It starts in the womb. Yes, it starts with good genetics. However, I have heard many accounts of women who state that during pregnancy, the child "pulled all the calcium from my teeth." Let us examine this further.

During pregnancy, the developing embryo needs calcium for developing long bones, and the source of its nutrition is its mother. There is no clear evidence that the enamel from the mother's teeth is dissolved by the embryo's need for calcium. If the calcium from the teeth dissolves, it enters the body through the throat into the stomach through the gastrointestinal system. The calcium then needs to be re-absorbed in the digestive tract, so it can enter into the mother's bloodstream and be available for the baby to absorb. If "the baby pulls calcium from my teeth," we would have to visualize calcium dissolving from the tooth and going directly into the bloodstream. I am not aware of a process where the unborn child pulls calcium from the teeth directly into the bloodstream and absorbs it through the placenta by way of the umbilical cord.

It is understandable that as the embryo extracts the calcium it needs from the bloodstream by way of the placenta, that calcium is replaced in the mother by her nutrition. In extreme cases, calcium can be withdrawn from the mother's long bones. The teeth are not long bones, as they are attached to the unique bones of the jaw by tiny ligaments called the periodontal ligaments. So if the baby pulled calcium from the teeth, it would have to come through these ligaments into the jawbone and into the bloodstream, or through the blood vessels inside the teeth, wearing down the inside of the teeth.

It is feasible that calcium could be drawn out of the upper and lower jaws to serve the developing embryo. However, the lower jaw (mandible) is like no other bone in the body. The dynamics of getting calcium out of this bone would be different for the upper jaw because it is not fixed. It has a diarthroidal joint, meaning it articulates in two areas with two articular disks, one on the right and one on the left. Most joints in the body meet at only one point so coordination is simplified. The lower jawbone has to coordinate its movement where movement in one joint affects movement in the other joint. The movements are coordinated with the muscles of mastication, the tongue, the lips, and the teeth, to produce the complex tasks of the mouth. There is no direct connection with the rest of the bones in the body. However, it is feasible that calcium can come from the bone marrow of the lower jaw into the bloodstream by way of the blood vessels.

The upper jaw (the maxilla) is connected directly to the skull and is generally made up of less dense bone than the bone in the lower jaw (mandible).

The mystery of the reports from mothers of the embryo pulling bone from their teeth is further compounded by the fact that the maxilla and mandible are very anatomically different, yet most women do not report more damage to their upper or lower teeth. There is an explanation for the oral symptoms reported during pregnancy, and I will shed some light on what happens.

There are estrogen receptors in the gums (gingiva) that respond to increased levels of estrogen by becoming more inflamed in the presence of plaque. In someone not performing proper hygiene, there will be an increased incidence of plaque buildup on their teeth and inflammation of the gums. In addition, feelings of nausea and the inability to maintain adequate nutrition can affect the mother's health. The pressure of the developing baby on the stomach sometimes leads to increased acid reflux, as well as regurgitation and vomiting, bringing acidic fluids into the mouth.

Acid becomes the major problem with pregnancy, because in tests with soda and diet, it has been shown that a more acidic environment leads to breakdown and dissolving of enamel in the teeth. When a patient has drying of the mouth in combination with dehydration from vomiting, higher estrogen levels, more plaque from swollen gums, plaque buildup, regurgitation, and vomiting it leads to increased acidity in the mouth. Acidity is the reason—along with improper oral hygiene—why some women lose enamel from teeth as well as speed up the decay of teeth that only had little or no decay before they became pregnant. For anyone considering pregnancy, a trip to the dentist to ensure proper dental health is recommended.

4

Transitioning from Childhood to Puberty

By age two, most of the twenty primary teeth are in place and the child should have been to the dentist at least once for an examination. Fluoride treatments will help to harden teeth against the ravages of children's diet and improper oral hygiene. Until about age seven, you should help your child to brush properly.

I recall one parent, whose child had multiple cavities at every visit, telling me, "Dr. Bonnick, I do everything right with this child. I brush her teeth every night myself. She does not eat candies. We have no candies in the house anymore. My husband and I have decided not to buy snacks. I just don't know what to do!"

"Well ..."

"Dr. Bonnick, it must be genetic. My whole family has soft teeth. It runs in the family. This dental thing is getting expensive, and I have three children. They all come here."

"How do you handle it in the morning after they eat breakfast?" I asked. "When do you brush their teeth in the morning?"

"Dr. Bonnick, we are in a mad rush, always in a mad rush to get them to school. We are always running late; the kids hold us up. Do you have kids? Doesn't everybody make their kids brush their teeth when they get up? Their breath could smell bad. We don't have time to brush after breakfast because we are rushing to get to school."

"When they get to school, do they brush?"

"They can't brush in school. The teachers won't allow them to brush in school. Aren't they supposed to brush when they first get up?"

"They should brush after eating to remove leftover food so the bacteria can starve."

"Why does everyone say they should brush when they get up?"

"I don't know where everyone gets their information, but as an expert in dental health, the rule I would give you is after any meal or snack or drink except water, all of us should brush our teeth."

"By the way, I know 'it is not cool to brush in school' but if possible, arrange for your children to brush while they are there. If not, you will be spending your money with me fixing cavities—and your child will start developing her ideas about the dentist from her repeat experiences."

"Dr. Bonnick, they don't brush when they come home. They snack. They drink a soda. They eat dinner and then do homework, or eat dinner then watch TV, and then brush their teeth before going to bed."

"Now, because they brush once before going to bed and brush essentially clean teeth again before eating breakfast, they are wearing plaque on their teeth for the rest of the day until bedtime. This is a prime scenario for cavities and gum disease to develop. This is costing us all in the long run."

"I'm not sure what to do. Dr. Bonnick, would you give me a note for the teacher?"

"Absolutely! If you have the children brush when they wake up, have them brush again after breakfast, and if they brush after lunch, they'll be on their way to better oral health. Promise them you will subtract the money for filling their cavities from the money you spend on birthdays, holidays, and vacations. Just kidding! Let's try better oral hygiene. Dentistry is not meant to be a punishment."

It is amazing how many well-intentioned parents and teachers will tell the children to brush their teeth two times per day. ***The proper way to care for teeth and preserve them from the destruction of cavities is to brush within twenty minutes after each meal, drink, or snack.***

As a dental student, I remember doing research for my senior project at the University of Maryland Dental School with a biochemistry professor. My job was to grow *Streptococcus mutans* (the most important bacteria to the creation of cavities in teeth) and study their adherence to smooth surfaces in different concentrations of sucrose similar to concentrations found in nature. The goal was to find out how these bacteria stick to the smooth part of teeth and enable other bacteria to stick to the tooth surface.

I obtained sugar cane juice from the government agency that quarantined it and found the sucrose concentration to be around 20 percent. I made up samples of 20 percent, 10 percent, 5 percent, and 2.5 percent sucrose by diluting the sample with distilled water and used control samples of the same concentrations drawn up with pure sucrose (cane sugar) and distilled water. My hypothesis assumed that adherence of the bacteria would be less with the natural juice than with pure sucrose. This unpublished study gave us two results: the bacteria had similar result to adherence between the pure sucrose and sugar cane juice, but surprisingly, the bacteria in the sample showed more adherences in the sample at 2.5 percent than at 20 percent. (The bacteria glued themselves to a smooth surface easier at a lower concentration of sugar.)

Application of this knowledge to the mouth shows us that activity of the bacteria is not that high at higher sugar concentrations, but as sugar in the mouth becomes more diluted with time, it promotes more adherences of *Streptoccocus mutans.* So if you had a fruit or some juice with sucrose and you did not brush right away, the lower concentration of sucrose (dilution) would make the bacteria more active (with time, bacteria sticks to teeth better) and cause more holes in your teeth.

Many parents are health-conscious, and sometimes it works to the detriment of their children. Some refuse to have their children drink fluoridated water, and instead they use bottled or filtered water. These children not only lose the benefit of the hardening properties of fluoride when their teeth are developing, but they should use topical fluoride more often at a time when they are more likely to have dental cavities.

At approximately six years of age, the first permanent teeth start appearing in the mouth, and it is very important to have them sealed to protect them from cavities at a point when children are not very good at cleaning their teeth. The baby teeth are often partially covered with gum tissue, and because it hurts to brush, some children neglect the proper cleansing of the area. The first large back tooth is called the first molar. This is one of the most important teeth for stabilizing the bite, and it usually appears when the child is six years old. It is the most important back tooth, and it represents the "sweet spot" for chewing. This is the tooth that is most often filled, extracted, or in need of root canal treatment because it is one of the first permanent teeth that appears in the child's mouth. Loss of the first molar is a common reason for posterior bite collapse (back teeth don't fit together well, so the bite is not as efficient as it could be), and if it is lost, immediate replacement with an implant is of extreme importance.

The bone grows with the eruption of adult teeth. It is therefore not a good idea to replace missing teeth with implants in the growing patient. Until the end of the "growth spurt," bone development could change the relationship of the crown of the tooth with the developing bone. If a first molar or any primary tooth is lost early, your child should be evaluated for a space maintainer (a device to maintain the space for an erupting tooth, prosthesis, or implant).

The mixed dentition stage is a confusing time for most parents. The primary teeth are being lost as permanent teeth are erupting into place and the child is going through a growth spurt. Puberty gingivitis is common as hormonal changes as well as poor flossing habits (non-flossing) by children promote excessive plaque buildup. Lack of flossing leads to more cavities between the teeth, so teeth that may be sealed on top can still have cavities through the sides. Root canal treatments, as well as large fillings, are common during this time.

Orthodontic treatments are started during this transitional time for many reasons. Growth spurts can be used to correct malocclusions, and loss of teeth allows the orthodontist to move teeth into more ideal positions. There are generally three classifications of occlusion (the way teeth come together). Class one occlusion is one in which the relationship of the upper and lower teeth is ideal. The upper teeth should be on the cheek side of the lower teeth and the teeth relate properly. Class two occlusion is a relationship where the upper jaw and teeth stick out too far from the lower teeth and the relationship is one of a protruding upper jaw. In a class three occlusion, the lower teeth stick out ahead of the upper teeth and give the appearance of a "strong jaw."

An overjet is not an abnormal condition, but when the distance between the upper and lower front teeth is so far apart that it makes incising (biting) difficult or it stresses the lower jaw to move extremely forward in order to touch the upper jaw, the overbite should be treated. An overbite is the measurement of the upper front teeth closure over the lower. When an overbite is extreme, the lower teeth sometimes hit into the upper gum near the roof of the mouth. This should be corrected. A crossbite can exist with one or more teeth when the relationship of the upper teeth is on the tongue side of the lower teeth. A slight overlap (degree) may not be harmful, but your dentist can determine whether it should be evaluated or not. A crossbite left untreated could result in facial deformities and improper development of one or both jawbones.

The major challenge during the mixed dentition stage is to get the child to brush after every meal and floss once or twice per day. The environment and parents' attitudes set the tone for early-adulthood habits and determine how much it will cost for dental care in the future.

5

Problems of Early Adulthood

This is the period when proper early care makes the biggest impact. Those who have not learned how to brush or floss properly will become concerned about their bleeding gums, bad breath, removal of unnecessary wisdom teeth, and their first bout with a real toothache. This is also the stage during which exploring or new activities lead to accidents requiring swift and expensive action to ensure proper preservation of the dentition. This period is also when those who had braces and who did not practice proper oral hygiene will require cosmetic work. The growth spurt is nearing its end or is complete, and placement of implants to replace missing teeth can start.

For many, this is the time when they find themselves in a place other than home for purposes of college, job, or relationships. At this point, many are trying to save money by not spending, and one of the things that are sometimes neglected is dental care. Some have skipped out on the extraction of wisdom teeth earlier and now find that they will have to take time off from their job or school to pursue extractions and recovery.

Those who had never had a bad dental experience and did not practice the habit of brushing after all meals and flossing twice per day will start to struggle with bad breath, gum disease, and cavities. Long hours on the job or at school make the bad habits worse. Treatment by the dentist for repeated bouts of teeth and mouth infections is commonplace. Those with good habits may try to improve their teeth by doing bleaching or cosmetic bonding.

Braces make performing proper hygiene more difficult, and I always recommend that during active orthodontic treatment, instead of cleanings twice per year, the patient receive cleanings four times per year. The savings of proper oral hygiene and need for less gum surgery afterward will more than pay for the two additional treatments per year. Breakdown of enamel around the brackets when proper, and additional, hygiene measures are not pursued, will be similar to the breakdown that occurs during pregnancy or from extreme soda use without proper and immediate brushing and flossing afterward.

Motorcycling, partying, skiing, contact sports of all types, seizures, and altercations can lead to the immediate destruction or loss of teeth. This will need immediate treatment to minimize the damage, and many young people may not have saved up an emergency fund for a situation like this. Many end up in later years spending tens of thousands of dollars to take care of defects that they were unable to take care of at the time of the emergency. Others will end up in dentures or temporary appliances of various types. These temporary measures can affect the social life of a young adult and decrease their self-confidence.

One of the rites of passage from being a dental student to being a dentist is practicing a procedure for the first time on your lab partner. We practiced impressions on each other. We practiced cavity detection on each other. We practiced injections on each other. One day, my lab partner was doing a periodontal (gum disease) evaluation on me.

"Bert! You have gum disease!"

"What?" I said.

"Gum disease. You have bleeding, lots of bleeding points. Have you ever been treated for gum disease?"

"No. How bad is it?"

"You have some pocketing, and bleeding from the pockets. Here, let me go over it with you."

I spent numerous visits in the clinic getting my gum disease treated and perfecting my brushing and flossing techniques. It was disheartening, and I felt lousy for neglecting my dental health for so long. I was happy that I did not have to go through the trauma of extractions again because of my neglect.

A bloody toothbrush without pain is often ignored, but that is a mistake. Gum disease starts slowly and painlessly. In smokers, it can be undetected because of the effect of nicotine in constricting blood flow; therefore, they may not see bleeding as readily as nonsmokers. Smokers also need more regular check-ups to detect oral cancers.

A traditional cleaning will not be sufficient to decrease gum disease. Additional treatment of scaling and root planing with local anesthesia and/or conscious sedation may be needed. Treatment may have to be repeated, or more advanced surgery will have to be done. More than two visits to the dentist per year may be needed to keep gum disease at bay. Sulcular (space between the tooth and the gum) and oral antibiotics as well as rinses may be needed, and in cases that do not respond properly, a visit with a periodontist (dental specialist in treating gum disease) is in order. Gum disease is the most common cause of bad breath.

Toothaches can occur for the first time during this transitional time. A "real toothache," as I refer to it, is one in which decay breaks down the teeth, to the point where bacteria get into the nerve and cause an irreversible pulpitis (inflammation of the tooth pulp). The options at this time do not include a filling; the nerve of the tooth has to be taken out or the tooth has to be extracted. The removal of the nerve and subsequent post buildup and then a crown usually takes care of the problem, but in the event that it does not, the dentist will usually do an extraction. If the extraction site is filled with bone-grafting material, it preserves the bone, and in several months a replacement tooth implant can be placed.

Altercations or on-the-job accidents can cause broken teeth, and the treatment usually requires some type of cosmetic restoration of the teeth. Some teeth may need to be extracted if the fracture line extends below the gum into the supporting bone. Other teeth may need root canal treatment if the fracture is just into the nerve of the tooth. An unsalvageable tooth can sometimes be treated with an immediate implant that can have a temporary done immediately after placement. Grafting helps to maintain or create new bone for placement of the implant or provide the shape of bone, so if a dental bridge is made, it does not have a large space to trap food.

Delivering a denture or removeable prosthesis to a patient for the first time can be very stressful for both doctor and patient. Sometimes a denture is all they can afford. I feel very sad when I have to remove a tooth that has no cavities because trauma has made it non-restorable. The last choice for a patient is to leave the dentist with the space for the missing tooth unoccupied.

I feel a sense of loss and a bit of remorse when I assist a patient in losing a part of their body—albeit a tooth. After all someone who loses a part of their body is starting a journey—this is the beginning of a long road that is going in the wrong direction.

Three Periods of Dental Health - The First Forty Years - The Second Forty Years - and The Third Forty Years

The first forty years is the period during which sacrifices have to be made for your children, and it is a time when your priorities become secondary to those whom you are responsible for taking care of. Some feel guilty when they take care of their dental problems during this period, because they think their resources should be spent on their dependents. During this period, emphasis should be on prevention.

Sometimes the prevention requires that you restore multiple teeth with crowns before they turn into more extensive work. This may be the period that requires that you wear a bite guard to protect your new crowns and to help prevent damage from clenching at night. If you have established good habits, early routine maintenance and an occasional crown or filling will be all you would need.

This is the time when you help your family to establish good habits and keep your family dental costs down. It is also the time when the *easy* solution is not always the *best* solution.

The second forty years put you in the position to do what you have always wanted to do for yourself. If you had poor dental care early and you have taken care of the children, now you have time to take care of yourself. It is during this time that many patients seriously evaluate their priorities and summon the courage to make their treatment decisions based on what will be the best for the long term.

Sometimes the missing tooth that you had ignored can be restored by an implant. We have to be prepared that the space cannot accommodate an implant or the bone is shrinking or the adjacent tooth has moved partially into the space. Sometimes the bone has shrunken to the point where nothing short of a major procedure like grafting some hip bone there can help you out. The consequences of the first forty years are coming back to help you or to give you regrets. Hopefully, you will never know the inconvenience of wearing a removable appliance.

You quickly rush to the bathroom after eating to try to get a seed from under your "plate," and you try to be cool about it when your friend comes into the bathroom and greets you. You lock the door so your partner never gets to see you without your plate, because you have to take the plate out to get the seed from under it. You cannot go to a new restaurant, because you are not sure you can properly chew what is on the menu.

The flip side is you never knew what it is like to be without your own teeth. You are not embarrassed by your smile. You eat nuts, salads, fruits, meats, or whatever you want. You don't have a denture container to carry around when you go on an overnight trip, and you are convinced that it is great to be alive and have your teeth.

The third forty years is a cautious time of your life. Your dental health has impacted your quality of life. Early in the 1900s, the Mayo Clinic found that most people who enjoy a long life have their own teeth. Back then they were not sure why, but now we know that dental disease can exacerbate other systemic diseases, and affects your quality of life.

I am reminded of the gentleman who came in at age ninety-five to have his teeth restored and told me the story of why he decided to do so. "At age eighty, I decided I did not have long to live, so I did not have my mouth fixed. Now it is fifteen years later, and I realized that I spent all that time in discomfort when I did not have to."

7

Implant Dentistry

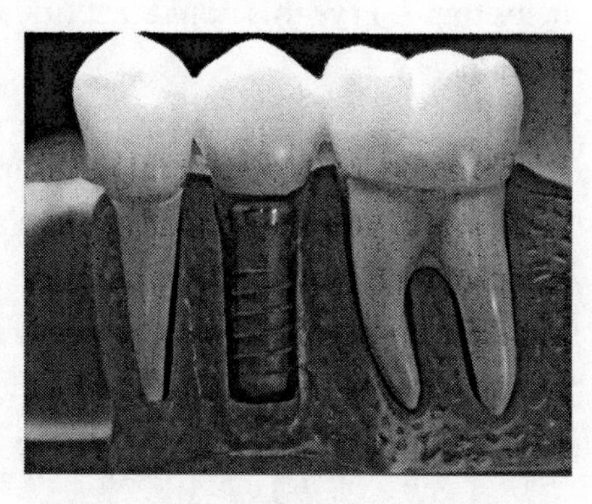

Dental implants come in many different forms. There are subperiosteal implants, ramus frames, blades, and root form implants. The subperiosteal implant is mostly used for areas where the bone is not of sufficient height to accommodate root form implants, and is mostly used in shrunken lower jaws. The ramus frame and blade implants are mostly used in the lower back jaw when the patient is not an ideal candidate for advanced bone grafting techniques or the doctor is proficient with this procedure and offers it as an option. The most common type of implants used most of the time are root form implants that mimic the root of the tooth. The reasons given in this chapter are addressed primarily to the root form implant.

The implant consists of a titanium alloy that is biocompatible to bone and encourages the bone to form around it, holding it in place. The healing process is referred to as osteointegration.

Once the implant becomes osteointegrated, it helps to preserve alveolar bone. Approximately 60 percent of the bone in your jaws grew in response to the erupting teeth. When a tooth is extracted, you go through a process of losing bone in that area that can last a lifetime. An implant helps to preserve this bone structure by slowing down the shrinkage.

It also helps to prevent malocclusion. When you lose a back tooth, you set up shifting in the other teeth. Teeth adjacent tend to move into the space, and the teeth opposing drift up or down into the newly formed space. Some spaces become “food traps” and contribute to bad breath and periodontal disease.

Implants can slow facial atrophy- shrinkage which occurs in facial muscles when they are not properly exercised. Roughly 150 newtons of force are applied on normally functioning back teeth, while only 50 newtons are applied on the front teeth. This force keeps the facial muscles toned and maintains our “looks”. If missing teeth are replaced with implants, this force is restored and good facial muscle tone is preserved. Additionally, the ability to chew properly vastly improves the digestion process.

Teeth are more efficient at chewing than dentures, sometimes two or three times more. There is an increased risk of aspiration of food if you have false teeth. Aspiration of food is one of the causes of death in the elderly that hardly gets mentioned, except in CPR courses.

In order to keep false teeth more stable, many denture users will place denture adhesive inside their denture to help keep them in place. In America, about $148 million is spent on denture adhesives every year.

Complete lower dentures usually move about ten to twelve millimeters during chewing and contribute to the need to chew softer foods that may be less nutritious. Many denture wearers develop anti-social tendencies because of their limited ability to chew a wide variety of foods. Many are unwilling or are uncomfortable when they go out for a meal. Dental implants will stabilize dentures, crowns, and bridges and allow for more efficient and confident chewing.

Many general dentists in the Unites States of America do not surgically place implants, although many of them advertise one of their services to be "dental implants." What they mean is they will plan and restore your implant after a specialist surgically placed the implant. The disadvantage I found when I did it this way twenty years ago is that if the patient had a problem with the implant, they did not have a direct recourse with one dentist. A concern for the patient is when the restoring doctor and the specialist disagrees on who should take responsibility for defective materials or accidental damage to the implant. I have found that most good dental teams are willing to help the patient restore their dentition by redoing the work at a reduced fee.

Note: It is the patient's responsibility to keep the information on the size and make of the implant in the event the patient relocates or their dentist(s) retire, leave the practice or change locations.

In the long run, implants are the most cost effective, and efficient option. Some patients choose bridgework to restore dentition because they lack the bone needed for an implant or because they think this is a less costly option.. However, when a dentist prepares one tooth for a crown, the chances of needing a root canal is about 3-5% in a five year period. When at least two teeth are prepared for a three-unit bridge it increases the odds of needing a root canal over the next five years from 3 percent to about 15 percent. The additional cost and the possibility of needing to redo your bridge after root canal treatment could end up costing you more than you thought you were saving over the cost of doing an implant. Cost is measured in money, but more costly is the inconvenience and time away from your other activities.

Did Grandma die early because she did not have implants?

In 1999 my grandmother passed away at the age of seventy-nine on the way to work. She was a firm believer that "If you are not sick, you should have a job." She also had a great sense of adventure and often accompanied me on my mission dental trips or visiting another state or country. Although we had arranged for her to live with my mom and dad, she insisted on "turning her own key in her own door."

Grandma, like a lot of those born in the 1920s, got her denture by the time she was forty years old, and like a lot of people, she did not like her bottom denture because it had no suction. I was not insistent on her having implants because she wanted to save her money to support her independence. In retrospect, this might not have been the best long-term decision, because not having implants meant she could not crush her food properly in order to receive the best nutrition. Studies show that dentures are about 30 percent efficient at crushing food, implant-supported dentures around 70 percent, and natural teeth around 90 percent.

Grandma complained about loss of appetite, stomach pains, and irregular bowel movements. She had medication that needed to be taken with meals to prevent further stomach irritation. On that fateful morning, her lack of appetite meant she did not eat in order to take her medication, and her increased blood pressure led to a stroke from which she did not regain consciousness.

I miss my grandmother and wish I could have done more for her. I have committed myself and my practice to the promotion of the benefits of having dental implants, because I think it can make a difference in people's lives. No amount of money saved was able to keep Grandma with us for additional years, but money properly invested in making sure she could eat more than soups and "mush" might have given me additional years with her. If you are missing some teeth and your dentist suggests implants, you should consider it a good option. Listen with an open mind. My Grandma would have wanted me to give you the option of implants, and I hope you agree.

Evaluating the Bone

As soon as a tooth is extracted, blood forms a clot in the socket and new bone cells migrate into the socket area, accompanied by immune cells and cells that form supporting tissues. Remodeling leads to the formation of new hard and soft tissue. At initial healing the largest amount of bone loss occurs at the extraction site, and then bone-loss slows but continues gradually over the years.

The jawbone is composed of alveolar and base bone. Alveolar bone forms in response to the erupting teeth and starts to go away when the tooth is lost. Base bone provides the foundation on which the alveolar bone develops.

When we look at the series of pictures of an edentulous jawbone (a jawbone without teeth) over time, we notice that bone loss progresses after loss of a tooth and can be accelerated when pressure is placed on the tissue by wearing a denture. In order to slow bone loss after extracting teeth, it is recommended that implants be placed, preferably after the extraction sockets are grafted.

To slow bone loss and the shrunken jaws associated with aging, people wearing any type of denture should be evaluated for implants. National standard of care recommends that if you wear a complete lower denture, you should have at least two implants for lateral stabilization and more if possible for bone preservation.

People are living longer, and they will need to preserve their oral structures longer. The loss of even one strategic tooth could make for a long and miserable "golden years." After all, what are the golden years like without the ability to chew, digest, and receive proper nutrition?

The same principle of implanting new teeth for accident victims or people born with congenitally missing teeth applies. Congenitally missing teeth is a condition where the patient has fewer than a full complement of teeth. With new advancements in implants, it is getting easier to find an implant to replace the missing teeth.

Mini-Implants

Any implant with a diameter less than three millimeters is considered a mini-implant. An implant needs to be surrounded on all sides by bone if it is to have the best chance of surviving over a long time. Historically, many implants were small because patients had lost bone as a result of extractions and/or wearing fixed or removable dentures. As the ability to grow (graft) new bone has increased, the need for smaller implants may decrease. Initially, smaller implants were considered temporary implants to stabilize temporary teeth until the larger implants had a chance to integrate with the bone. The mini-implants were then removed and new dentures attached to the larger implants.

Today the technology for mini-implants is so good, we are finding that you can place the implant and on the same day attach the patient denture crown or bridge and have a high degree of success. Several years ago, I started to use the best of both worlds. I would place mini-implants at the same time I placed normal implants and attach the denture to the minis while letting the larger implants integrate. In a lot of the cases we completed, we found that the aggressive threads of the minis that allowed you to use them right away, along with the new technology and surfaces, made it difficult to remove the minis, so we incorporated them into the final prostheses (dentures).

If a patient had a well-fitting lower denture that moved during chewing and did not want to wait four to six months until their implants integrated, they could have an initial evaluation and the necessary diagnostic X-rays and/or CT scan so they could be "sized" based on the remaining jawbone for implants. On the day of the surgery, they could be pretty confident that they would leave with an implanted denture with some degree of stability.

The Sterngold Company, a longtime manufacturer of attachments for dentures and crowns, came up with their version of using mini-implants with conventional implants adapted to work with their ERA (Extra-coronal Resilient Attachment) system that allows the prosthesis a small range of movement and enhances their longevity. A rigid attachment puts a lot more force on the implants than a resilient attachment and being able to replace the resilient part of the attachment is an advantage of the ERA system. This system has since been bought by the Zimmer dental company. Implant Direct also manufactures a similar system that already has the attachment for the popular Locator attachment.

If you could see a series of models of the lower jawbone just before and after teeth were lost, you would see a consistent pattern. Initially, the jawbone is at its maximum height and width, and supports the patient's facial features very well. When a tooth is extracted, the alveolar bone that grew in response to the erupting tooth starts to shrink also. When the first denture is made, it is usually the best, because most of the alveolar bone is there for support; however, as bone loss continues, the denture sinks further into the soft tissue and the height of the lower face contracts, giving the patient an aged appearance. The next denture should be made larger, to compensate for the shrinking bone. But when the dentist does this, the patient usually complains, because they had slowly over time adapted to the first denture. At that point, the dentist usually cuts back the new denture so it could feel like the first denture, except it does not feel as secure, because some of the bone was lost. This bone loss continues until the nerve of the jawbone is close to the surface and causes the denture wearer discomfort. At that point, the patient usually suffers with one denture after another, looking for a dentist who could give them the feel of their first denture.

The way to slow this bone loss and aged look is to place socket grafts and/or implants as soon after extraction as is possible, because alveolar bone stays around implants as if it was the root of a tooth. In effect, implants not only help to retain dentures but help to maintain bone as well.

What Are Combination Cases?

Our office prides itself in our ability to take care of multiple problems in a reasonable period of time, with or without sedation. Over the last few years, we had several cases that involved sinus surgery. With our current technologies, we are able to assess the amount of bone that exists in the upper jaw before we encounter the membrane of the sinus. The sinus membrane is very forgiving and can repair itself in six weeks. If there is enough bone, seven millimeters or more, we can use blunt instruments to lift up the membrane like a blanket and insert bone grafting before placing the implant. If the amount of bone left is seven millimeters or less, I may elect a two-stage technique.

In the two-stage technique, bone is grafted below the sinus by displacing the membrane upward through a small hole made in the side of the bone. The membrane is then protected by a collagen liner and bone graft is packed, leading to a new bone growth of ten to twenty millimeters, or more. Six months later, implants of varying lengths are placed and allowed to integrate as the grafted bone continues to mature. After integration of the implants, implant-supported teeth are placed. The patient does not go without teeth, because they wear a transitional prosthesis (temporary dentures or bridges) while they wait on the body to heal.

In our office, we use several systems of implants, depending on the results we are trying to achieve. Sometimes we have to do extractions, bone grafts, root canals, fillings, or gum treatments; place implants; or place transitional appliances all during the same visit. If the case warrants it, we can use mini-implants to hold initial appliances securely until the long-term implants heal. A lot of prior planning goes into providing these treatments for combination cases.

Some patients who live a long distance from the office or even in another state can have a CT scan made and the results e-mailed to us, so we can do the virtual surgery and have all the supplies needed to perform the procedures before they arrive at our office.

Cosmetic Dentistry

For as long as I have been a dentist, I have been taking classes and teaching other dentists the value of cosmetic dentistry. On completing my second year of post-graduate training in hospital dentistry in 1988, I became an instructor, and I was given the job of solving cosmetic problems for our patients in our hospital general practice. Some of these patients had teeth broken after intubation for general surgery. My job was not only to give them new teeth but to make them look good. During those years, I learned the importance of working with various dental specialists, as well as knowing what was needed in order to get the final results. The doctors doing the treatment were new dental graduates in their first years of practice. Since the doctors were graduates of dental schools from all around the United States, they sometimes presented different ways of solving the same problem. Needless to say, I learned just as much as I taught in that program.

Tooth-colored fillings are what we call *composites.* These have replaced acrylics as long term esthetic filling materials, because they are better able to match the color of your teeth and last longer. Composites used in dentistry are basically hydrocarbons with quartz and glass particle fillers. Composites are widely used in the aerospace and other industries to provide for lighter strong materials that can be formed into various shapes. Porcelain is the other long-term material that we use for crowns and bridges, which are mostly tooth-colored.

In the past, porcelain fused to metal or all-metal crowns was the technology for crowning teeth. This technology has served us well and is still the treatment of choice in cases where you have decay or fillings with margins that go below the gums. Porcelain fused to metal is also the technology used to build porcelain bathtubs.

With the advent of Cad Cam technology, and zirconium use as a substructure to porcelain, we now have many options to make teeth look beautiful. 3M® Company uses technology to color the zirconium substructure the same color as the root structure, and then lab technicians apply porcelain to it, creating beautiful teeth. The resulting crown maintains the tooth color over a long time and is very esthetic. It costs a little more, but it is well worth the investment.

Cad Cam technology uses a camera to take a picture of the prepared tooth and fabricate the missing part of the tooth, using computer correlation technology to mill it. In-office Cad Cam machines can be found in some offices and are mainly used to restore posterior (back) teeth. Most cosmetic anterior aesthetic crowns or veneers are still finished by the artistic skills of trained lab technicians. Some technicians use the Cad Cam technology to fabricate the substructure and then apply porcelain for the final result.

A patient came to my office with an interesting request. "Hey, Doc, I want a smile just like yours," she said.

"Do you mean the shade of the teeth or the shape?" I asked.

"I just like the look of your smile."

I completed my examination and found she had no upper teeth and only the lower front teeth. The remaining bone in the upper was enough to support an upper denture, but on the lower, I would have to place some crowns and a lower appliance.

"Do you have any pictures of you in your late teens or early twenties? One where you have a full smile?"

"Yes! And yes, I would like to smile like that again."

I collected all the information and proceeded to make an upper smile that showed teeth when she smiled. The contour on the appliance was shaped like real gums, and the contours of her teeth were made like her own teeth from the pictures. No metal clips showed on her lower appliance, and we later took pictures together that showed that indeed she could smile as well as I did.

Cosmetic dentistry involves smile creation, and the materials are not always porcelain veneers. All restorations can be used as cosmetic enhancements. The reason I encourage the use of bone grafts after extractions is that it creates the contours that we need to enhance the cosmetic result. All dentistry that is considered functional can also be cosmetic.

Is Cosmetic Dentistry Art or Science?

Dentists have long known that different styles of teeth fit people with different facial structures. We have used the concepts of the Golden Proportion to help us with rebuilding an aesthetic and functioning smile. Dentists who are very serious about cosmetic dentistry benefit from joining and participating in the various programs set up by the American Academy of Cosmetic Dentistry. The more an office combines the science of materials and techniques with the artistic talents of sculpting a harmonious smile, the closer they get to great results and a satisfied patient.

As a dental resident at Long Island Jewish Medical Center, I clearly remember a young lab technician, who worked on the development of thin porcelain wafers to cover unattractive teeth, telling us how they could solve cosmetic problems. They were called porcelain veneers. I became one of the early adopters of this conservative treatment mode with my longest case that I know of, still in place since 1989. Later on, I studied with Larry Rosenthal, who gave hands-on courses for other dentists. His techniques on straightening rotated teeth were very helpful then and are still helpful today. Good fundamentals are always helpful. In North Carolina, Ross Nash and Rob Lowe have passed on valuable information to dentists, from the use of photography to lasers used to enhance your results.

Cosmetic dentistry may involve working with different dental specialists to get a cosmetic, as well as functioning, occlusion. New types of braces, those that are "invisible" as well as "internal," can provide answers to adults who may be self-conscious during treatment.

The most important part of your cosmetic treatment is that one dentist is in charge of coordinating all the treatment that leads to the result that the patient wants. Sometimes part or parts of your treatment have to be done by another dental specialist, and your cosmetic dentist will do the final restorations.

A periodontist may be needed to treat your gum disease or sculpt your gums so that when you smile, the amount of gum shown is not unattractive. An oral surgeon may be needed if you have lost considerable amount of bone, need your jawbone repositioned or impacted teeth extracted. A prosthodontist may be needed if you have had extensive tissue loss from cancer treatment or an accident requiring a special prosthesis. An oral and maxillofacial radiologist may be needed if a 3-D model shows an unidentified artifact. An endodontist may be needed if you require a root canal treatment, or retreatment of a previous root canal, to make your cosmetic restoration feasible.

Most of the cases treated in our office require combination treatments, with or without sedation. Sometimes we have to do an extraction, place an implant, do a bone graft, root canal a tooth, place a fiber post, and then choose the right type of porcelain for each tooth to get the smile to look uniform. A lot of patients ask for veneers or lumineers; however, they may not be a candidate for those options because several teeth require several other types of treatment. The more a patient knows about what they want, the more we can advise them if it is achievable or unrealistic. Sometimes the achievable requires a little more perseverance on the part of the patient and doctor. Cosmetic dentistry is a combination of science and art.

9

Dentistry and the Heart, Artificial Joints, Chemotherapy, and Radiation

My youngest sister was born with a heart defect. She had repeated visits to her physicians for monitoring and did well enough through high school and into early adulthood. Her diagnosis was mitral valve prolapse, and the doctors thought that heart valve replacement would give her a better quality of life. My parents wrestled with the decision, but like most caring parents had a hard time signing the consent form that had death as one of the possible complications of surgery.

When my sister turned sixteen and I opened my first dental office, she worked with me as a part-time dental assistant. When she turned eighteen she was still being followed by the cardiologist at a local hospital, and made a decision that she would not have surgery unless it was unavoidable.

At twenty-two she was living at home and continuing her college education. On a day I will never forget, my receptionist told me my mom was on the phone.

“Bertrand it’s your sister!” Mom started.

“What’s happening?” I queried.

“They’re working on her. The EMS.”

“Is she alright?”

“She’s gone, I can’t…”

“Where are they taking her?”

“Kings County Hospital.”

“Mom, I’ll meet you there.”

My mother’s voice was in a tone that I never heard before or since. It was sure the result was not going to be good. I rushed back and forth, told my receptionist to cancel all remaining appointments, and rushed to the hospital.

At the hospital the last sight I saw of my sister was her body passing by on a gurney as one Emergency Medical Technician was performing chest compressions.

Several days later my Dad and I went to identify her body in the morgue. We held onto each other for support, and in silence while being choked up we wobbled to the car together. It was over for this young child, and it was uncertainty for the young two year old child she left behind. We could not believe it.

The death certificate stated as cause of death "Floppy mitral valve." The rest of the family was subjected to ECG, Echocardiograms, and Stress tests. Needless to say, my concern for patients with heart problems increased to the point where I took the courses given by the American Heart Association and became Advanced Cardiac Life Support Certified.

The connection between the heart and the mouth has long been established, and anyone with a heart murmur or artificial joint knows that they may have to take antibiotics prior to dental treatment. Antibiotic prophylaxis is medication taken in high enough doses to treat bacteria from the mouth that gets into the bloodstream to prevent it from damaging the heart or artificial joints. The mouth, throat, and sinuses are places that are conducive to the breeding of viruses, bacteria, and fungus that can spread throughout your body if your immune system is compromised.

Chemotherapy can compromise your immune system to the point where your remaining white blood cells are not enough to keep the body from coming down with a general infection. The general guidelines that your dental team gives you for taking care of your teeth and gums should be strictly followed to ensure that if you ever find yourself in the position where you need Chemotherapy--you will not suffer severe oral symptoms.

Radiation treatments affect the body in many ways. The bones in the jaws are particularly susceptible to radiation because the radiation compromises the blood flow in the bones. The salivary glands are adversely affected if they become exposed to radiation. The salivary glands are scarred and generally reduce their production of saliva causing dry mouth. Saliva serves a protective and lubricating function in the mouth, but dry mouth leads to multiple oral problems including increased cavities and gum ulcerations. The worse condition that can occur after radiation treatment is osteoradionecrosis. Osteoradionecrosis is the breakdown of the bones that happens after radiation treatment and can be exacerbated by ongoing oral infections. Hospitals treating this condition often use the hyperbaric chamber where patients are subjected pressurized oxygen over several appointments. The remaining jawbone that survives after treatment, presents major problems for restoring teeth because it is hard to make a satisfactory prostheses for a compromised jawbone.

My best advice is to take prevention of gum disease and cavities seriously. Make sure your oral health is in the best of shape before doing any elective surgery, or prior to the need for chemotherapy, or radiation treatment. So let us review the April 2007 recommendations by the American Heart Association.

Increased use of antibiotics also increase the risk of an allergic reaction to a specific antibiotic and that can be even more immediately life threatening than bacteria in the bloodstream. Consult your physician if you are uncertain about needing antibiotics prior to dental treatment. The following are basic guidelines for needing antibiotics prior to treatment.

1. Prosthetic cardiac valve
2. Previous bouts of infective endocarditis
3. Certain congenital diseases (Unrepaired cyanotic Cardiac Heart Defects (CHD), first six months after prosthetic repair of CHD, repaired CHD with residual defects, palliative shunts and conduits, etc)
4. Cardiac transplant recipients who develop valvulitis
5. Mitral valve prolapse with associated defects that your physician thinks may put you at risk for endocarditis.

   In addition to the above, certain cardiac surgical procedures may require that you wait six months prior to having a dental procedure, providing you do not have a life-threatening dental infection. If there is any uncertainty the best option is to see your physician prior to seeing the dentist.

Over the last few years more scientific studies have come out supporting the link between poor oral health and cardiovascular diseases. Current advances in DNA studies will allow doctors to predict the patients who are more at risk for diseases than others based on oral salivary or blood samples. Some dentists are using screening tests for C-Reactive proteins and early diabetic screening to see which patients may require close monitoring of their dental health.

Many studies have indicated the correlation between gum disease and cardiovascular disease. One in office treatment involves the use of an antibiotic delivered directly into the space between tooth and gum where severe periodontal disease needs to be treated. The area is cleaned and irrigated before the antibiotic is administered. The protocol for using this method can be discussed with your dentist and dental hygienist. Our Registered Dental Hygienists (RDHs) are trained to administer the treatment in conjunction with the dentist's diagnosis. Therapy can be repeated in areas of your gums that are infected, and referral to a periodontist will be made, if indicated.

10

Toothpaste, Mouthwash, and Fluoridated Water

Fluoride has been shown to harden enamel when it is incorporated into the forming of the enamel matrix. The more it was studied, the more the realization came that excess fluoride would cause "mottling" of teeth, and excessively large amounts of fluoride has been shown to contribute to weak bones in the body as it gets incorporated into the long bones. I support the American Dental Association's (ADA) position that if fluoride at the concentration of one part per million is placed in the water supply, the benefit for the population involved will be a lowering of the cavity rate.

Some people have called the move to fluoridate drinking water a plan to "poison the water and promote health problems." The other objection is that the fluoride used is the by-product of aluminum production and therefore it is contaminated and should not be used. Free fluoride ions are free radicals, and they can combine with hydrocarbons and produce fluorinated hydrocarbons. Some anti-fluoride advocates insist that these compounds are cancer-causing, and that is their reason for opposing fluoride.

The health benefits of having a chlorinated water supply have been shown to outweigh the risks by keeping the levels of pathogens low. The benefits of fluoride have been shown by numerous studies pre- and post-fluoridation, and without fluoride, we would need to increase the number of dentists to deal with the rampant cavities that would exist in our population.

In the original fluoride findings, communities with naturally fluoridated water sources showed significantly lower cavity rates. The studies from the fluorosis allowed us to come up with an optimal level of one part per million, which was enough to give the positive benefits without the negative effects of mottling or discoloration, or excessive fluoride buildup in the bones. If subsequent water supplies used a fluoridation process that allowed us to have optimal levels of fluoride and the method of placement in the water supply was transparent, I believe the public would benefit from it.

If communities are adding pure fluoride and are doing sampling to make sure some parts of their communities do not have excess fluoride in their water by pooling or improper mixing, then the benefits could ensue. If you are unsure about your water supply or are using well water, a water test would help you determine the existing fluoride level before you determine if your children need fluoride supplements. Some well water is naturally fluoridated, so make sure it is checked.

Some parents opt to use bottled water to avoid fluoridated water, and they experience the increased dental expense from having their children's teeth filled or needing root canal treatment. So if your community can provide you the source of their fluoride (for your peace of mind) and you follow up with water testing, you should feel good about enjoying the benefits of fluoride.

Another way of enjoying the benefits of fluoride is through topical fluorides and use of fluoridated toothpaste. Neutral fluorides are a good way to avoid the doubts of excess free radicals and fluoridation techniques, but caution should be taken not to swallow a lot of fluoride paste, as that can cause a stomach-ache or, in large amounts, fluoride poisoning.

Brushing with a fluoride paste can be followed by expectorating (spitting) without rinsing and waiting half an hour to eat or drink. Topical fluoride can be absorbed into the tooth structure to harden it and become a deterrent to cavity formation. Table salt is a neutral compound made up of a positive sodium ion and two negative chlorine ions; in like manner, fluoride ions are neutral in neutral fluoride compounds that are in topical fluorides and toothpaste.

I have never been surveyed by a toothpaste company, but I know in their ads they always talk about how many dentists recommend their brand of toothpaste. I recommend most fluoride toothpastes for everyday use, but recommend toothpaste without sodium lauryl sulfate for the first few days after the patient has a bone graft procedure. Sodium lauryl sulfate breaks down salivary pellicle, making your teeth feel smooth after you brush; it also retards new epithelial tissue development at a critical time when your newly grafted tissue needs to stay covered.

The new formulations of toothpaste promise whitening, tartar reduction, and fresher breath. The truth is all these benefits are available by doing your recare visits every three months. If the toothpastes did what they promise, there would be no need for dentists. Plain fluoride toothpaste with no addition of newer—sometimes harsher—chemicals, and regular dental visits seems to be the answer for most of the population.

Speaking of harsh, many mouthwashes have alcohol as one of their active ingredients, but the tide seems to be shifting to newer, "healthier" options that are nonalcoholic. One of the best mouth rinses for periodontal disease is one that contains chlorhexidine gluconate. It should be short-term therapy only because it stains teeth, especially in areas where bacteria and plaque accumulate. This mouth rinse is dispensed at the pharmacy and generally requires a prescription.

11

Osteoporosis and Your Dental Health

As the population ages, we tend to see more cases of osteoporosis: decreased bone density. Both men and women are at risk for developing this condition which makes them more susceptible to fractures from everyday activity. It is possible for someone suffering from osteoporosis to break their hipbones in such a way that compromises internal organs, and can lead to death. Any bone that breaks from the application of force presents a challenge to the patient, especially in an individual who enjoys an increased lifespan.

Men and women enjoy their maximum bone density in their twenties and early thirties—after which it can all go downhill. Weight-bearing exercises, proper nutrition, and medications have helped us battle the inevitable ravages of aging. The genetic factors are beyond our control at this point, but maintaining a good dentition is a big part of fighting the battle against osteoporosis.

Some bone cells build up bones (osteoblasts), while others remove old bone (osteoclasts), and the dynamic actions of both cells promote bone health. The slowing down of the bone-building process initially leads to osteopenia and then osteoporosis.

Early loss of bones in the jaws is initially caused by losing teeth, usually followed by wearing dentures. Teeth and implants provide weight-bearing stimulus to the jawbone and enhance bone deposition where it is most needed. Early bone loss associated with the loss of teeth is not osteoporosis, but it can occur at the same time with osteoporosis.

Regular dental care allows us to preserve teeth and bone, which helps with our nutrition. In the event of a health issue like osteoporosis, better oral health means we have to deal with fewer complications. Women seem to be at a higher risk for osteoporosis because they live longer, have more hormonal variations to contend with, and traditionally are involved in fewer weight-bearing exercises.

Medications like Boniva, Fosamax, Reclast, as well as hormonal and nutritional supplements, have been used to combat osteoporosis. Complications exist with all forms of therapy, but of particular interest to dentists are complications from bisphosphonates-related Osteonecrosis of the jawbones (BRONJ).

Risk factors for BRONJ include ulcerations under dentures, infections from periodontal disease or cavities, and trauma. Eliminating these risk factors decreases the likelihood of developing BRONJ. The incidence of jawbone necrosis increases if a patient has received IV bisphosphonates and later develops infections in an area of the jawbone that incorporated the bisphosphonates (studies suggests lower incidences with oral bisphosphonates). Trauma or surgery affecting the drug enriched areas of the jawbone can also start the breakdown of bone that leads to BRONJ. Some patients have ended up losing parts of their jaws while trying to increase their bone density with bisphosphonates because of oral infections.

If you have good dental maintenance and you need to be treated for osteopenia or osteoporosis, make sure you have all dental work taken care of prior to taking bisphosphonates. Once you start therapy, it may be too late to do major dental treatment or prevent the adverse effects of infected teeth or gums on your supporting alveolar bone.

12

TMJ, Headaches, and Facial Pains

(Special note to patients, this subject can be very complicated and is meant to inform you of the complexity when trying to self-diagnose. The help of a professional is seriously advised in this area. Consult your dental professional for further advice.)

TMJ is not a medical condition; it is a joint that articulates the lower jaw to the base of the skull. The part of the lower jaw that articulates with a disk that articulates with the skull is called the condyle. There is a disk between the rounded joint and the base of the

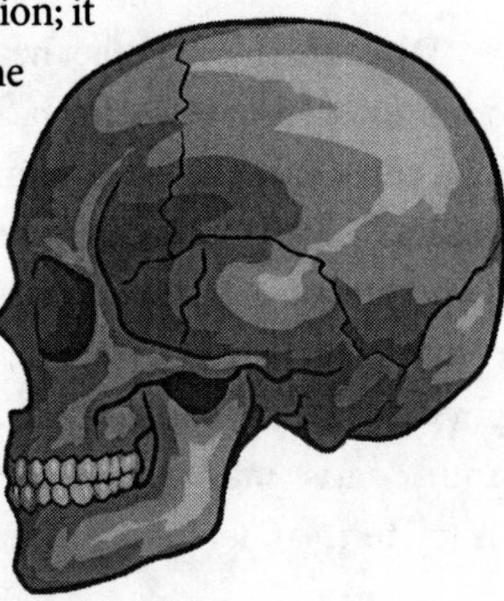

skull, so unless you have a worn disk there will not be a direct bone to bone contact. The condyles can come under extreme stress when the mouth is crushing food, the patient is clenching or grinding, or because of oral habits like nail-biting, or chewing gum, and trauma. A small muscle attaches in the front part of the disk and pulls it forward when the lower jaw translates forward, and damage to the disk or slipping of the disk compromises the area between the muscle and the disk and causes pain.

Any disorder to the condyles, disk, or muscles supporting chewing is called temporomandibular disorder (TMD), and this is the correct way to talk about problems in this area. TMD can strike at any time in life but seems to show up after a traumatic event or during a time of increased stress. Close to 90 percent of TMD is a result of muscle spasms and can be corrected with physical therapy, bite splints, and medications. The remaining disorders may have to be treated surgically or may continue as part of a systemic condition such as arthritis.

TMD can also trigger other headaches as your body recruits other muscles to move the head and neck. Headaches can also come from referred pain from areas called "trigger points." Headaches can be caused by infections, high blood pressure, trauma, bouts of migraine, tumors, teeth coming together in unfavorable ways (malocclusion), and physiological processes that are still being investigated. Your dentist may request tests, refer you to a specialist, or try to treat dental conditions that may be aggravating your condition. When there is pain in the head and neck region of non-dental origin, care should be taken to rule out the presence of tumors.

Facial pain refers to any of the conditions that cause abnormal and painful conditions in the head and neck region. There are dentists who limit their practice to this area, and they are very well informed and can diagnose and treat many conditions. There are many medical specialists who specialize in the treatment of chronic pain, and they use many forms of therapy to help patients.

Chronic pain often has a psychological component, so referral to a professional for counseling is not uncommon when treating facial pain.

13

Root Canals, Apicoectomies, and Extractions

At first glance, a tooth may seem like a piece of hard bone that chews food. However, if cut in cross section, a healthy tooth has a network of nerves, arteries, and veins that respond to stimulus from the outside as well as from the inside. Repeated drilling or cavities on a tooth allow oral fluids, water, chemicals, and bacteria to find their way into channels called "dentinal tubules," and they could eventually affect the nerve inside the tooth. Sometimes the effect will manifest years after the initial insult.

Pulpitis is the inflammation of the pulp (inner part of the tooth bearing the blood vessels and nerve). The condition of the pulp can be described using three words: reversible, irreversible, and necrotic. Reversible pulpitis refers to a fleeting and low-grade inflammation sometimes caused by stimulation or sensitivity of the tooth. Irreversible pulpitis

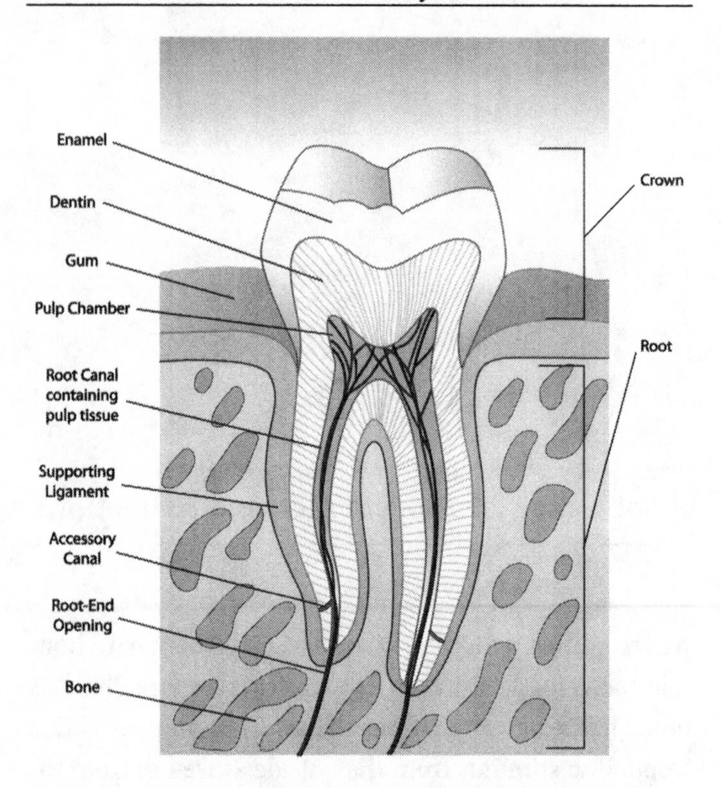

means the nerve is damaged or infected to the point that the tooth will not recover unless pulpal treatment is performed. Necrotic pulp means damaged, infected pulpal tissue that may no longer be responsive to stimuli or is infected. Root canal treatment is usually reserved for irreversible pulpitis and necrotic teeth.

The root canal process involves preparing the tooth with instruments a little deeper than a large filling, in order to locate the pulp and identify the canals. Instruments like files and reamers, driven by motor or used by hand can remove the pulpal tissues. Solutions that clean and disinfect are used to irrigate the canals as the residue is suctioned away. The length of the roots is determined with the help of diagnostic equipment as well as radiographs (X-rays), and eventually if the clinician determines, the canals can be dried and filled with root-canal cement along with a non-metallic filler There are numerous materials that can be used, and some can be partially removed after the filling of the root to make a space for a cemented post to support filling material and a crown. Materials and instruments are constantly being developed to make this treatment faster and easier for the patient.

Intentional root canals are beneficial in several cases. Sometimes the tooth is in a rotated position, and a cosmetic as well as functional solution would mean reducing the area of the tooth that sticks out from the ideal arch form. When removal of the tooth structure will lead to encroachment on the nerve, an intentional root canal will ensure that the patient will not have to suffer undue pain later. Clinical judgment is also used when we approach close enough to the nerve that putting in a filling may irritate the nerves and cause the patient undue suffering.

There are people who think that root canal treatment should not be done, because there are always residual bacteria in the complex network of the nerves of a tooth that can lead to future infection and immune response. I say the sooner problems are intercepted, the better the chances of having a good outcome.

Root canal (endodontic) therapy requires special skills and intimate knowledge of the structure of teeth. Most front teeth have one canal; other teeth may have two, three, four, or more. Some lower front teeth have complex canal systems that make it difficult to adequately find and clean them out. Calcified canals result when the pulp chamber lays down calcium salts on the inside of the canal in response to an insult in order to protect itself. Sometimes these canals are hard to find and may require multiple visits or referral to a root-canal specialist (endodontist).

Some studies indicate that root canal treatment can have a 93 percent success rate up to five years after the treatment was initiated. The other 7 percent of teeth may need to be retreated, extracted, or have an apicoectomy performed. An abscess at the root of a tooth that had a root canal, a post placed and then a crown, may not lend itself readily to being retreated. However, the gums can be lifted off the bone and the abscess at the root removed. When the tip of the infected root is removed, a filling is placed in the remaining root to seal it from being re-infected. This is called an apicoectomy, where the apex is the end of the root of the tooth. When a root canal fails or a tooth fractures vertically or horizontally below the level of the bone, the best option is to take the tooth out, remove the infection, place a bone graft in the area, and then place and restore an implant at a later date.

Extractions of teeth that had root canal treatments present some special problems, because the tooth is more brittle and may break easily. This problem is common when a patient does not follow up after their root canal by having the tooth crowned. The brittle tooth breaks and the tooth needs to be extracted. Cavities can also destroy a root canal tooth if the patient does not discover it in time; cavities do not cause a root canal tooth to hurt and patients who do not see their dentist often may miss the fact that a cavity was developing.

14

Current Dental Developments

A prospective patient called my North Carolina dental office from Washington, DC, inquiring about the cost of five implants. I sent her to a CAT scan (CT) center to obtain a scan. After it was sent to me via e-mail, I used a dental software called Simplant® to determine the size and density of the available bone and tried several implants, virtually, in the area the implants would be placed. My treatment coordinator contacted the patient, and two weeks later, she was in my office, getting ready for her implants!

Another inquiry came from an office three hours away. The patient went to the CT center; I downloaded the CT scan, did the virtual surgery on the computer, and then e-mailed the completed plan to Belgium for production of surgical guides. Along with the surgical guides, they sent me a model of her mandible produced from the computer files. I actually used the model to practice the surgery, where the objective was to place four 8-mm implants into 11mm of available bone. The surgery was completed according to plan, and the patient was finally able to have a denture that stayed in placed because it was anchored by implants.

Three-dimensional dental scans can also detect calcifications of the carotid artery or thyroid tumors, as well as sinus problems. They can also show complex root canals and the extent of infection at the root of the tooth.

Another recent development in dental technology is the use of Lasers. Lasers can be used to help treat gum disease, fill cavities without anesthesia, or help treat root canals. They may even be used to perform gingivectomies which remove sufficient gum to give the effect of longer teeth.

New scans can make an impression of your teeth and have it e-mailed to the lab, where the scan is turned into models. This avoids messy impression materials and keeps germs away from lab personnel.

The days of dark-colored (amalgam) fillings are almost over, because the materials that match the tooth color are getting so much better. No longer do dentists have to be bothered by rules that tell them to dispose of the remnants of amalgam fillings in a covered jar with dental fixer because the mercury vapors will be harmful to your employees. I agree with the American Dental Association and The Academy of General Dentistry as they point out in their studies that no harm has come to anyone with amalgam fillings, except for those who may be allergic to one of their components (silver, nickel, tin, zinc, copper, etc.). Many states have started to require that offices have amalgam separators to keep the materials—especially mercury—out of the water system. Hazardous-material disposal services will properly dispose of amalgam in an environmentally safe way. We will still be removing millions of amalgam fillings for years to come, because they have served us well.

Ozone therapy, as well as pharmaceutical formulations, is being used to regenerate enamel in teeth. Genetic engineering of tooth materials has been studied to come up with alternatives to repair teeth.

Finally, dental offices are referring patients to other physicians as their screening examinations become more exacting. Saliva, as well as fluids in the space between teeth and gum, can be used to detect various illnesses. Early bone or tooth loss can indicate problems with white blood cells, and many more developments will benefit patients.

15

Dentures: The Last Resort

"I will just pull my teeth, and that will be the end of my teeth problems," is one mantra that I used to hear a lot. When teeth go, bone and supporting tissues will go also; then we have to deal with prosthesis (fixed or removable bridgework). The bone in which teeth erupt is called *alveolar bone;* the supporting bones of the jaws we call *base bone*. If you keep your teeth intact throughout your lifetime, you will preserve most of your jawbones. As you lose teeth, the alveolar bone also shrinks, and it shrinks even more if you have a denture putting force on it.

Along with loss of jawbone, you have movement of teeth into the now-empty space. Fewer teeth mean more collapse of the bite and an older, more "collapsed look." Generally, the first denture will fit better than subsequent dentures because there is more bone support. Since most people do not strictly adhere to removal of denture for six to eight hours per day and replacement of dentures every five to eight years, they will wear away the jawbone even more.

Lower dentures are notorious for moving and should be anchored by implants to preserve bone and to prevent movement of the denture. Many denture wearers spend millions of dollars for denture adhesives. Some denture adhesives have a high concentration of zinc, and because they use more than the recommended dose, wearers get more zinc in their system. Zinc competes with copper for absorption and depletes the amount of copper in the body. Low levels of copper can lead to neurological problems, and there have been incidences where cases of paralysis have been tracked to copper deficiency. Denture adhesive companies are developing new formulations to reduce or eliminate zinc.

Most denture wearers will not let you know, but the efficiency of chewing using a set of full dentures is about 35 percent, while normal-functioning teeth are over 90 percent. Nutrition and longevity suffers also, because food that is not properly crushed is more difficult to digest, and the diet of most denture-wearing patients contains little in the way of raw leafy vegetables and nuts. This translates into suffering for most denture wearers. Nutrition suffers. Neurological function suffers. Proper functioning of the gastrointestinal system suffers. Social interactions suffer. Physical beauty suffers and longevity suffers.

More than ninety years ago, the Mayo Clinic, in their study of longevity, found a positive correlation with natural teeth and length of life. Better nutrition, better health, and better appearance are all a result of taking care of natural teeth. Loss of appetite, a drying of the mouth from medication and aging, and poorer health are all a result of poor dental hygiene and maintenance.

Other special problems with dentures include decrease in taste, change in salivation, allergies to acrylic, sore spots, bone shrinkage, and risk of nutritional deficiencies. Health issues increase, shattering the perception that dental health is not important. Bad breath and dehydration are part of the world of denture wearing.

My best advice for denture wearers is to keep as many teeth as possible; use implants where you can to retain dentures and maintain the alveolar bones of the face.

16

The Best Dental Insurance—Saving Money through Prevention

The term "dental insurance" is not an exacting term, a more accurate term would be "dental benefits". Misunderstandings about the expected insurance payments, and the patient's responsibility for copayments, are a potential source of conflict between dental staff and patients. Dental insurance is not insurance in the traditional way we expect insurance to work. If you bought an insurance policy for your house and it burns to the ground, you would expect your insurance to cover the rebuilding of an equivalent house. If your tooth broke, you would expect "dental insurance" to pay for its replacement.

If a patient wants some dental service and they have what is considered "good insurance", and expect the majority of their fees to be covered, the conversation may go like this.

"How much does my insurance pay for a crown?"

"Since this is major work, it would be fifty percent of the UCR minus your $50 deductible."

UCR (Usual, Customary and Reasonable) fees are determined by the insurance company and it can be misunderstood by the patient. The fact of the matter is that the UCR for a particular zip code is different for different insurance companies, and so it is clear that they are not using the same statistics to determine UCR.

The insurance company will not tell you how they determine what a crown should cost in your zip code. They are a for-profit business and they have to pay out less than they receive in order to remain solvent. The benefits that a patient receives are related to actuarial tables and how much money the insurance company collected for premiums. If you are part of a PPO (Preferred Provider Organization) you may receive a schedule of services that the plan will consider for reimbursement. You have no recourse if a preventive service is not covered; you will have to pay for it. If you go to a dentist who is not signed up with the network, you will be considered "out of network," and the consequences will be lower reimbursement.

Patients who understand this model should understand that choosing the best treatment for the diagnosis is their responsibility and they have to decide their course of treatment even if costs them more.

Direct reimbursement plans repay patients for dental work they have already paid for, using a formula that allows patients to direct where they spend their money. The freedom patients have to choose their own dentist, and how the reimbursement is allocated, has increased the popularity of this type of plan. The American Dental Association has provided information for its membership to encourage patients to solicit their employers to consider this type of plan.

Dental benefit plans that allow you to use any dentist will also have a fee schedule, but you will not necessarily know all of the reimbursement fees unless you do a predetermination of benefits. Many states have insurance commissioners who will support you in your right to receive a benefit if you were entitled, even if there was no predetermination.

Some people use the excuse "I don't have dental insurance" as the reason they do not take care of their teeth. The correct term that represents the dental plans is "dental benefits."

When your employer shops for benefits, they compare premiums and consider benefits. Some benefits are sold as PPOs (Preferred Provider Organization), where dentists sign up to accept decreased fees in return for a listing in the plan and more patients. The decreased fees are usually about 20 percent less than UCR (usual customary reasonable) and may require the dental office to cut overhead in some way to maintain a reasonable profit. There are many good dentists who accept PPOs as a service to their patients. The contracts are very strict; some services are not covered, and the dentists may charge their full fees if their patients want a service that is not covered. The contracts are long term and the dilemma for me, as a dentist when I accepted these plans, came when the patient wanted what the insurance paid for, and I knew that was not the best treatment for their diagnosis.

Patients should choose their doctor deliberately and not because "I have no choice; they accept my plan."

In the early days of dental insurance "But you take my insurance" meant "Please do your billing so I do not have to pay anything out of pocket." That is unethical. When the insurance company sets the fee that the doctor will accept and only pays a part of that fee, the patient must pay the balance. If the fee for a root canal set by the insurance company is $800 and they pay $640 minus $50 deductible, meaning a net payment of $590, the patient is responsible for paying the $210. Patients nowadays understand how insurance benefits are paid and set their budgets for dental services accordingly.

MOUTH-TO-MOUTH-SPREAD GERMS

As DNA studies have improved, so has our understanding of the disease processes that affects oral structures.

Some of the pathogens that cause the most severe gum diseases can be passed from one family member to another through direct or indirect means: directly by kissing or indirectly by sharing utensils. The kind act of a mother cooling soup to feed to her infant child or tasting the food before she feeds her child could transfer infectuous pathogens that lead to advanced disease (eg. A.A. Actinomycoses Actinomycetemcommitans a major pathogen for periodontal disease). The patient's dental care and oral hygiene program, or lack thereof, influences how well the germs take root in the space between tooth and gum.

Leftover food between the teeth and gum, plus the warmth of the mouth, allows pathogens to thrive. The immune response the body sets up, along with the by-products produced and excreted, leads to a breakdown of bone and supporting tissues. Loose teeth, bad breath, exposed roots of the teeth, and eventually tooth loss results. The inability to properly chew food affects our ability to maximize our nutrition.

Currently, we know that bacteria from gum disease and infection from rotten teeth can get into a patient's bloodstream and cause heart problems; Chronic gum infection in a pregnant woman can lead to the premature birth of a child; I have referred patients to their physicians for screening for diabetes based my observation of their early onset of infection, and difficulty treating their gum disease. Healing can be slowed if the patient is an undiagnosed diabetic, and control of diabetes is sometimes enhanced, and easier, when there is decreased infection in the body.

A young child in Washington, DC lost his life when the abscess from his tooth precipitated a brain abscess and led to his untimely death. Death from a rotten tooth is a horrible way to go, and could have been easily prevented.

If you paid extra money for four preventive visits to the dentist per year, most of your major dental problems would be detected early, and they would be less costly to take care of. There are times when worn teeth will need to be crowned—another way of replacing worn enamel. You may choose to improve the function and concurrently improve the looks of your teeth.

There are large nerves that run directly from the oral structure to the brain without traveling through the spinal cord. They are so sensitive that you can detect a strand of hair in your mouth or a small fishbone. Children use their mouths to interact with things they come in contact with; it is one of the ways they explore the world. When you are stressed, it even manifests in your oral structures.

In the coming years, DNA will provide clues about the people who are more susceptible to oral and other diseases. Currently, dental offices provide screening for oral cancer, high blood pressure, diabetes, and other physical ailments. There are even in-office tests for the presence of harmful bacteria and the presence of inflammation in your body that can help guide your dentist in determining the best treatment for you. Nutritional awareness and training has been a part of the dental curriculum and will assume more importance as more research became available. The best is yet to come from dentistry.

In 1840, the first dental college in the world was founded in Baltimore, Maryland. The Baltimore College of Dental Surgery University of Maryland Dental School was formed because of the vision of two men: Horace Hayden and Chapin Harris. Hayden and Harris were unsuccessful in convincing the medical school to train dental specialists, like some of their counterparts in Europe, so they took the matter to the legislature. The Maryland state legislature persisted, voted to allow the formation of the dental school, and dentistry became an independent profession.

"God is from everlasting to everlasting" and "energy cannot be created or destroyed" are statements that lead me to believe that science cannot help but acknowledge God!

Bertrand Aristotle Andrew Bonnick
1986 Graduate of the Baltimore College of Dental Surgery.

For more information on how dentistry has changed people's lives, visit our website at **www.relaxdds.com**.

CPSIA information can be obtained at www.ICGtesting.com
Printed in the USA
LVOW08s0228091113

360319LV00004B/3/P